Android Development –

Frequently asked Interview Q & A

I0016031

By Bandana Ojha

Introduction

Android is a trend that has driven the overall advancement of touch-screen smart phones. More and more smart phones in the market today are using Android systems that will undoubtedly benefit millions of mobile users. Compared to the limited iPhone and iPad lineup, Android represents choice on a grand scale. As Android provides an interesting revenue model, a lot of organizations have a dedicated team for application development with hiring catching pace. This book contains most frequently asked android development interview questions and answers. It will help the reader to get a good understanding of methodology, concept, approaches and design patterns of Android application .This book is aimed at anyone who is interested to take a job in android development ranging from junior to expert level. The authors of this book conducted so many interviews at various companies and meticulously collected the most effective questions with simple, straightforward explanations. Rather than going through comprehensive, textbook-sized reference guides, this book includes only the information required for android development to start their career as an android developer. Answers of all the questions are short and to the point. We assure that you will get here the 90% frequently asked interview questions and answers.

1. What is difference between Serializable and Parcelable ?

Serializable is a standard Java interface. You simply mark a class Serializable by implementing the interface, and Java will automatically serialize it in certain situations.

Parcelable is an Android specific interface where you implement the serialization yourself. It was created to be far more efficient than Serializable, and to get around some problems with the default Java serialization scheme.

2. What is a Layout in Android?

A Layout in Android is a representation of a UI component, contained in items such a Fragments or a Widget. A Layout can either be contained in an XML file, or it can be instantiated at runtime (View and ViewGroup can be created and manipulated programmatically.

3. What is ADB?

ADB stands for Android Debug Bridge. It is a command line tool that is used to communicate with the emulator instance. ADB can control your device over USB from a computer, copy files back and forth, install and uninstall apps, run shell commands, and more.

4. What are ADB components?

ADB is a client-server program that includes three components:
A client, which runs on your development machine. You can invoke a client from a shell by issuing an ADB command. Other Android tools such as DDMS also create

ADB clients.

A server, which runs as a background process on your development machine. The server manages communication between the client and the ADB daemon running on an emulator or device.

A daemon, which runs as a background process on each emulator or device instance.

5. What is a fragment? Why they were introduced?

Fragments were introduced in the version 3.0 of Android to deal with new devices coming into the market (tablets). They provide a modular mechanism to create UI that can be easily reused throughout an application. As the previous versions of Android did only allow to have one running activity , it was not possible to make usage of the entire screen and present different modules. With this new approach, several fragments could be embedded in an activity , making an application much more interactive.

6.What are the different states of the activity lifecycle?

Once an activity is launched, it goes through a lifecycle.

The activity Lifecyle has four states.

Running:-When the activity is on the foreground of the application, it is the running activity. Only one activity can be in the running state at a given time.

Paused:-If the activity loses focus but remains visible (because a smaller activity appears on top), the activity is paused.

Stopped:-If the activity is completely covered by another running activity, the original activity is stopped.

Destroyed:-While the activity is paused or stopped, the system can destroy it if it needs to reclaim memory

7. What is dependency injection?

Dependency Injection is a design pattern to implement inversion of control, and to resolve dependencies. Dependency Injection (DI) eliminates boilerplate code and provides a much cleaner and effective code.

8. What are Intents?

Intents displays notification messages to the user from within the Android enabled device. It can be used to alert the user of a particular state that occurred. Users can be made to respond to intents.

9. Describe three common use cases for using an Intent.

Common use cases for using an Intent include:

To start an activity: You can start a new instance of an Activity by passing an Intent to startActivity() method.

To start a service: You can start a service to perform a one-time operation (such as download a file) by passing an Intent to startService().

To deliver a broadcast: You can deliver a broadcast to other apps by passing an Intent to sendBroadcast(), sendOrderedBroadcast(), or sendStickyBroadcast().

10. Is intent used to provide data to a ContentProvider?

Intent object is a common mechanism for starting new activity and transferring data from one activity to another. However, you cannot start a ContentProvider using an Intent.

When you want to access data in a ContentProvider, you must use the ContentResolver object in your application's Context to communicate with the provider as a client. The ContentResolver object communicates with the provider object, an instance of a class that implements ContentProvider. The provider object receives data requests from clients, performs the requested action, and returns the results.

11. What is an explicit Intent?

Android Explicit intent specifies the component to be invoked from activity. In other words, we can call another activity in android by explicit intent.

12. What is an implicit Intent?

Implicit Intent doesn't specify the component. In such case, intent provides information of available components provided by the system that is to be invoked.

13. What is a Sticky Intent in android?

Sticky Intent is also a type of intent which allows the communication between a function and a service.

14. What is the function of an intent filter?

As every component needs to indicate which intents they can respond to, intent filters are used to filter out intents that these components are willing to receive. One or more intent filters are possible, depending on the services and activities that is going to make use of it.

15. What is a broadcast receiver?

A BroadCastReceiver is a component that delivers a message system-wide, so any other component can capture it and interact with it. There are many of them provided from Android OS, such as broadcast to indicate a low level of battery, a disconnection of the Wi-Fi or network or a screen that just turned on.

16. What is sleep mode in android?

Sleep mode mean CPU will be sleeping, and it doesn't accept any commands from android device except Radio interface layer and alarm.

17. What is AIDL?

AIDL, or Android Interface Definition Language, handles the interface requirements between a client and a service so both can communicate at the same level through inter process communication or IPC. This process involves breaking down objects into primitives that Android can understand. This part is required simply because a process cannot access the memory of the other process.

18. What are the steps involved in creating a bound service through Android Interface Definition Language (AIDL)?

Following are the steps involved in creating a bound service through Android Interface Definition Language

-Define an AIDL interface in an .aidl file.

-Save this file in the src/ directory of the application hosting the Activity and any other application that needs to bind to this service — the latter is particularly important and is often overlooked.

-Build your application. Android SDK tools will then generate an IBinder interface file in your gen directory.

-Implement this interface, by extending the generated Binder interface and implementing the methods inherited from the .aidl file.

-Extend Service and override onBind() to return your implementation of the Stub class.

19. What data types are supported by AIDL?

AIDL has support for the following data types:
-string

-charSequence

-List

-Map

-all native Java data types like int,long, char and Boolean

20. What is a loader?

Loader and LoaderManager classes where introduced in Android 3.0 to remove workload from the main thread. A Loader performs an asynchronous loading of data in an Activity or a Fragment . They are continuously monitoring the source of their data, and when it changes they deliver the new data.

21. What is Android compatibility?

The Android Compatibility defines the technical details of the Android platform and provides tools for OEMs to ensure developer applications run on a variety of devices.

22. Describe how an OutOfMemoryError happens in android?

As the application progresses, more objects get created and heap space is expanded to accommodate new objects. The virtual machine runs the garbage collector periodically to reclaim memory back from dead objects. The VM expands the Heap in Java somewhere near to the maximum heap size, and if there is no more memory left for creating

new object in java heap, it will throw a java.lang.OutOfMemoryError killing the application. Before throwing OutOfMemoryError the VM tries to run the garbage collector to free any available space but if even after that there is still not much space available on Heap in Java, it will result into an OutOfMemoryError .

23. What are launch modes?

A launch mode is the way in which a new instance of an activity is to be associated with the current task.

24. What is Android Framework?

Android Framework is an important aspect of the Android Architecture. In Android framework you can find all the classes and methods that developers would need to write applications on the Android environment.

25. Why a Handler is used for?

You use Handler to communicate between threads, most commonly to pass an action from a background thread to Android's main thread.

26. Can you deploy executable JARs on Android? Which packaging is supported by Android?

No, Android platform does not support JAR deployments. Applications are packed into Android Package (.apk) using Android Asset Packaging Tool (AAPT) and then deployed onto Android platform. Google provides Android Development Tools for Eclipse that can be used to generate Android Package.

27. What's the difference between onCreate() and onStart()?

The onCreate() method is called once during the Activity lifecycle, either when the application starts, or when the Activity has been destroyed and then recreated, for example during a configuration change.

The onStart() method is called whenever the Activity becomes visible to the user, typically after onCreate() or onRestart().

28. What is Orientation?

Orientation decides if the LinearLayout should be presented in row wise or column wise fashion.

- The values are set using setOrientation()

- The values can be horizontal or vertical

29. What are the notifications available in android?

Toast Notification – It will show a pop-up message on the surface of the window

Status Bar Notification – It will show notifications on status bar

30. What is an activity in android?

An Activity can be understood as a screen that performs a very particular action in an android application. If you want to do any operations, you can do with activity.

31. Is it possible to create an activity in Android without a user interface ?

Yes, an activity can be created without any user interface. These activities are treated as abstract activities.

32. Enumerate three key loops when monitoring an activity?

Entire lifetime – activity happens between onCreate and onDestroy
Visible lifetime – activity happens between onStart and onStop
Foreground lifetime – activity happens between onResume and onPause

33. When is the best time to kill a foreground activity?

The foreground activity, being the most important among the other states, is only killed or terminated as a last resort, especially if it is already consuming too much memory. When a memory paging state has been reached by a foreground activity, then it is killed so that the user interface can retain its responsiveness to the user.

34. How to launch an activity in android?

Using with intent, we can launch an activity.

35. What is activityCreator?

An activityCreator is the initial step for creation of a new Android project. - It consists of a shell script that is used to create new file system structure required for writing codes in Android IDE.

36. What is Android?

Android is a mobile operating system developed by Google, based on a modified version of the Linux kernel and other open source software and designed for smartphones , tablets., Android TV, Android Watch, PCs and Android Auto.

37. Describe Android application Architecture?

Android application architecture has the following components-

Services – It will perform background functionalities

Intent – It will perform the inter connection between activities and the data passing mechanism

Resource Externalization – strings and graphics

Notification – light,sound,icon,notification,dialog box and toast

Content Providers – It will share the data between applications

38. What are the key components in android architecture?

The key components in android architecture are

Linux Kernel

Libraries

Android Framework

Android applications.

39. What are the advantages of android?

Open-source: It means no license, distribution and development fee.

Platform-independent: It supports windows, mac and Linux platforms.

Supports various technologies: It supports camera, Bluetooth, Wi-Fi, speech, EDGE etc.

Highly optimized virtual machine: Android uses highly optimized virtual machine for mobile devices, called DVM (Dalvik Virtual Machine).

40. What is the Google Android SDK?

The Google Android SDK is a toolset which is used by developers to write apps on Android enabled devices. It contains a graphical interface that emulates an Android driven handheld environment and allow them to test and debug their codes.

41. What is the importance of having an emulator within the Android environment?

An emulator lets developers play around an interface that acts as if it were an actual mobile device. They can write and test codes, and even debug. Emulators are a safe place for testing codes especially if it is in the early design phase.

42. Does android support other language than java?

Yes, android app can be developed in C/C++ using android NDK (Native Development Kit). It makes the performance faster. It should be used with android SDK.

43. What are the advantages of having an emulator within the Android environment?

The emulator allows the developers to work around an interface which acts as if it were an actual mobile device. - They can write, test and debug the code. - They are safe for testing the code in early design phase

44. What is Mono for Android?

A. Mono for Android is a software development kit that allows developers to use the C# language to create mobile applications for Android-based devices. Mono for Android exposes two sets of APIs, the core .NET APIs that C#

developers are familiar with as well as a C# binding to Android's native APIs exposed through the Mono.Android.* namespace. You can use Mono for Android to develop applications that are distributed through the Android Application Stores or to deploy software to your personal hardware or the Android simulator.

45. What is an Adapter?

An Adapter acts as this bridge and responsible for converting each data entry into a View that can then be added to the AdapterView.

46. What do you mean by a drawable folder in Android?

In Android, a drawable folder is compiled visual resource that can use as a background, banners, icons, splash screen etc.

47. What is portable wi-fi hotspot?

Portable Wi-Fi Hotspot allows you to share your mobile internet connection to another wireless device. For example, using your Android-powered phone as a Wi-Fi Hotspot, you can use your laptop to connect to the Internet using that access point.

48. What is a view in android?

A View in Android is the most basic element to build a UI element. A View is contained in the package android.view.view. They occupy a rectangular area in the screen, and they capture interaction events with the view.

They also render in the screen the content anytime it has been updated.

49. What is the support library, and why was introduced?

To deal with the Fragmentation among different versions of Android, the Support Library was introduced first in 2011. The support library package contains a set of libraries to provide backwards compatibility with previous versions of Android.

50. What are application Widgets in android?

App Widgets are miniature application views that can embedded in other applications (such as the Home screen) and receive periodic updates. These views have referred to as Widgets in the user interface, and you can publish one with an App Widget provider.

51. What is the difference between a regular bitmap and a nine-patch image?

In general, a Nine-patch image allows resizing that can be used as background or other image size requirements for the target device. The Nine-patch refers to the way you can resize the image: 4 corners that are unscaled, 4 edges that are scaled in 1 axis, and the middle one that can be scaled into both axes.

52. . How can two Android applications share same Linux user ID and share same VM?

The applications must sign with the same certificate to share same Linux user ID and share same VM.

53. What do containers hold?

Containers hold objects and widgets in a specified arrangement. - They can also hold labels, fields, buttons, or child containers.

54. 14. Why cannot you run standard Java bytecode on Android?

Android uses Dalvik Virtual Machine (DVM) which requires a special bytecode. First, we must convert Java class files into Dalvik Executable files using an Android tool called "dx". In normal circumstances, developers will not be using this tool directly and build tools will care for the generation of DVM compatible files

55. What is SMP?

SMP stands for Symmetric Multi-Processor. It describes an architecture for multiple processors accessing memory.

Android was supporting a unique processor architecture until Android 3.0. Most of the Android devices have different cores, so it makes sense to make use of them (even if they are natively prepared to run applications only in one processor and use the other ones for secondary tasks). Android provides a set of do and don'ts, such as not abusing volatile or synchronized variables.

56. What is the DDMS, and what can you do with it?

DDMS stands for Dalvik Debug Monitor Server. It is a tool used for debugging that can

perform multiple actions:

- It provides the possibility to capture screenshots from the device.

- There is an analyzer of active Threads and of the Heap state, very

helpful to deal with memory issues and leaks.

- It can simulate locations, incoming SMS or incoming calls, to deal with

Activity recreation.

- There is an integrated file explorer.

57. How would you upload multiple files to an http server in a single http request?

Use MIME Multipart. It was thought to send different fragments of information in a single request and is integrated with most of the native HTTP clients.

58. What is the importance of settings permissions in app development?

Permissions allow certain restrictions to be imposed primarily to protect data and code. Without these, codes could be compromised, resulting to defects in functionality.

59. What is AAPT?

AAPT is short for Android Asset Packaging Tool. This tool provides developers with the ability to deal with zip-compatible archives, which includes creating, extracting as well as viewing its contents.

60. What are ViewGroups ?

A ViewGroup extends from the View class. A ViewGroup is the base class used to create Layouts, which are containers for different sets of Views (or also ViewGroups).

61. Is it okay to change the name of an application after its deployment?

It is not recommended to change the application name after its deployment because this action may break some functionality.

62. What is the name of database used in android?

SQLite: An opensource and lightweight relational database for mobile devices.

63. Can you dynamically load code in android?

Although not recommend, code can be loaded dynamically from outside the application APK by making use of the class DexClassLoader.

64. What is a FrameLayout? When to use it?

A FrameLayout is a special type of view in Android that is used to block an area of the screen, to display a single element (very often used to display and position Fragments on the screen). It can however also accept multiple children. The items added to the FrameLayout are put in a stack, with the most recently added element at the top. If there are several children, the dimension of the FrameLayout will be the dimension of its bigger child (plus any padding if it was originally added to the FrameLayout).

65. What is the Role of SQLite Database in Android?

SQLite is an open source and lightweight database administration system that takes a small amount of disk storage, so it is an excellent choice of developers to develop a new application for Android. It is the first choice of developers because of the following reasons

Zero configuration database

Don't have server

Open source

Single file database

66. What are the different storage methods in android?

Android offers several different options for data persistence. Shared Preferences – Store private primitive data in key-value pairs. This sometimes gets limited as it offers only key value pairs. You cannot save your own java types.

Internal Storage – Store private data on the device

External Storage – Store public data on the shared external storage SQLite Databases – Store structured data in a private database. You can define many numbers of tables and can store data like another RDBMS.

67. How many ways data stored in Android?

1.SharedPreferences

2.Internal Storage

3.External Storage

4.SQLite Database

5.Network connection

68. . What is shared preferences in android?

Shared preferences are the simplest mechanism to store the data in XML documents.

69. What is nine-patch images tool in Android?

We can change bitmap images in nine sections as four corners, four edges and an axis.

70. What is application Widgets in Android?

Application widgets are miniature application views that can be embedded in other applications and receive periodic updates.

71. What is a service in android?

The Service is like as an activity to do background functionalities without UI interaction.

72.What is a content provider in android?

A content provider component supplies data from one application to others on request. Such requests are handled by the methods of the ContentResolver class. A content provider can use different ways to store its data and the data can be stored in a database, in files, or even over a network.

Dialogue Notification – It is an activity related notification.

73.What is container in android?

The container holds objects, widgets, labels, fields, icons, buttons, etc.

74. Where are lay out details placed? Why?

Layout details are placed in XML files - XML-based layouts provide a consistent and standard means of setting GUI definition format.

75. What is singleton class in android?

A class which can create only an object, that object can be share able to all other classes.

76. Why can't you run java byte code on Android?

Android uses DVM (Dalvik Virtual Machine) rather using JVM(Java Virtual Machine), if we want, we can get access to .jar file as a library.

77. What is an APK format?

APK is a short form stands for Android Packaging Key. It is a compressed key with classes, UIs, supportive assets and manifest. All files are compressed to a single file is called APK.

78. What is the API profile exposed by Mono for Android?

 Mono for Android uses the same API profile for the core libraries as MonoTouch. MonoTouch and Mono for Android both support a Silverlight-based API, without Silverlight's UI libraries (e.g. no XML, no WindowsBase.dll, etc.), and free of the sandboxing limitations of Silverlight.

79. What is the difference between Service and IntentService?

Service is the base class for Android services that can be extended to create any service. A class that directly extends Serviceruns on the main thread, so it will block the UI and should therefore either be used only for short tasks or should make use of other threads for longer tasks.

IntentService is a subclass of Service that handles asynchronous requests (expressed as Intents) on demand. Clients send requests through startService(Intent) calls.

80. How can ANR be prevented?

One technique that prevents the Android system from concluding a code that has been responsive for a long period of time is to create a child thread. Within the child thread, most of the actual tasks of the codes can be placed, so that the main thread runs with minimal periods of unresponsive time.

81. How will you pass data to sub-activities?

We can use Bundles to pass data to sub-activities. There are like HashMap that and take trivial data types. These Bundles transport information from one Activity to another

82. Name some exceptions in android?

Inflate Exception

Surface.OutOfResourceException

SurfaceHolder.BadSurfaceTypeException

WindowManager.BadTokenException

83. How many dialog boxes do support in android?

Following are dialog boxes supported in android

AlertDialog

ProgressDialog

DatePickerDialog,

TimePickerDialog

84. Define the application resource file in android?

JSON,XML , bitmap etc. are application resources. You can inject these files to build process and can load them from the code.

85. How to get a handle on Audio Stream for a call in Android? Permissions.PROCESS_OUTGOING_CALLS: Allows an application to monitor, modify, or abort outgoing calls

86. What is BroadReceivers?

BroadcastReceiver is a component that does nothing but receive and react to broadcast announcements.

87. When is the onStop() method invoked?

A call to onStop method happens when an activity is no longer visible to the user, either because another activity has taken over or if in front of that activity.

88. What is .dex extension?

Android programs are compiled into .dex (Dalvik Executable) files, which are in turn zipped into a single .apk file on the device. .dex files can be created by automatically translating compiled applications written in the Java programming language.

89. What is the Guardian app for Android?

Guardian app for Android delivers all the best content from guardian.co.uk to your phone or tablet. Read the latest news, sport, comment and reviews, watch video, listen to broadcasts and browse stunning picture galleries while on the move.

90. What features does it have?

Navigate by section, topic or contributor - Download your homepage and favorites for offline reading with the touch of a button, or schedule a daily download for a time that suits you - Browse our award-winning audio and video content - Save contributors, topics and sections to your favorites folder - Add favorites to your home screen with an expanded view or link - Swipe through stunning full-screen picture galleries - Share articles and galleries via the Android share function - View content in portrait or landscape orientation

91. Name various types of android applications?

1.Foreground

2.Background

3.Intermittent

4.Widget

92. Which dialog boxes are supported by android?

Android supports 4 dialog boxes:

AlertDialog: Alert dialog box supports 0 to 3 buttons and a list of selectable elements which includes check boxes and radio buttons.

ProgressDialog: This dialog box is an extension of AlertDialog and supports adding buttons. It displays a progress wheel or bar.

DatePickerDialog: The user can select the date using this dialog box. *TimePickerDialog:* The user can select the time using this dialog box.

93. What is a pending intent?

A pending intent is like an Intent . It is given to a foreign or third-party application and provides them for permission to execute a piece of code in an application. It is used very often which classes such as NotificationManager , AlarmManager or AppWidgetManager .

94. What is Android SDK?

To develop a mobile application, Android developers require some tools and this requirement is satisfied by "Android SDK" which is a set of tools that are used for developing or writing apps.

It has a Graphical User Interface which emulates the Android environment. This emulator acts as an actual mobile device on which the developers write their code and then debug/test the same code to check if anything is wrong.

95. Which components are necessary for a New Android project?

Whenever a new Android project is created, the below components are required:

manifest: It contains xml file.

build/: It contains build output.

src/: It contains the code and resource files.

res/: It contains bitmap images, UI Strings and XML Layout i.e. all non-code resources.

assets/: It contains a file which should be compiled into a .apk file.

96. Name the important core components of Android.

The core components of Android operating systems are:

Activity

Intents

Services

Content Provider

Fragment

97. Explain Activity Lifecycle.

When a user interacts with the app and moves here and there, out of the app, returns to the app, etc. During all this process "Activity" instances also move in the different stages in their lifecycle.

There are seven different states like – onCreate(), onStart(), onRestart(), onResume(), onPause(), onStop(), and onDestroy(). These are termed as a 'callback'. Android system invokes these callbacks to know that the state has been changed.

98. What is Orientation?

Orientation is the key feature in Smartphones nowadays. It can rotate the screen between Horizontal or Vertical mode.

Android supports two types of screen Orientations as mentioned below:

Portrait: When your device is vertically aligned.

Landscape: When your device is horizontally aligned.

setOrientation() is a method using which you can set a screen alignment. HORIZONTAL and VERTICAL are two values which can be set in the setOrientation() method. Whenever there is a change in the display orientation i.e. from Horizontal to Vertical or vice versa then onCreate() method of the Activity gets fired.

Basically, when the orientation of the Android mobile device gets changed then the current activity gets destroyed and then the same activity is recreated in the new display orientation. Android developers define the orientation in the AndroidManifest.xml file.

99. Which tools are used for debugging on the Android platform?

There are different tools for debugging which include – Android DDMS, Android Debug Bridge, iOS simulator, Debugging from Eclipse with ADT, Remote debugging on Android with Chrome etc.

100. How do you find memory leaks in the mobile app on Android platform?

Android Studio is using Android Device Manager (ADM), this ADM is used to detect the memory leaks in the Android platform.

When you open ADM in the Android Studio then on the left-hand side of the ADM, you will find your device or emulator in which a heap sign will be displayed. When you are running any mobile app then you will see the heap size, memory analysis and other statistics displayed on it.

101. What is the difference between a fragment and an activity?

An activity is typically a single, focused operation that a user can perform (such as dial a number, take a picture, send an email, view a map, etc.).

Activity implementations can optionally make use of the Fragment class for purposes such as producing more modular code, building more sophisticated user interfaces for larger screens, helping scale applications between small and large screens, and so on. Multiple fragments can be combined within a single activity and, conversely, the same fragment can often be reused across multiple activities. This structure is largely intended to foster code reuse and facilitate economies of scale.

A fragment is essentially a modular section of an activity, with its own lifecycle and input events, and which can be added or removed at will. It is important to remember, though, that a fragment's lifecycle is directly affected by its host activity's lifecycle; i.e., when the activity is paused, so are all fragments in it, and when the activity is destroyed, so are all its fragments.

102. Can you create custom views? How?

To extend an Android View and create our own one, we need to create a class that inherits from the View class, and that has at least a constructor that receives an object Context and another one AttributeSet , as follows:

class ExampleView extends View

{

public ExampleView (Context context, AttributeSet attrs)

{

super(context, attrs);

}

}

If you want to create custom attributes for a custom View , you need to declare a <declare-styleable> resource element and add it the custom attributes you need.

103.When does onResume() method called?

onResume() method is an activity lifecycle method. This is called when the activity come to foreground. You can override this method in your activity to execute code when activity is started, restarted or comes to foreground.

104. How can two Android applications share same Linux user ID and share same VM?

The applications must sign with the same certificate in order to share same Linux user ID and share same VM.

105. What are the downsides of Android?

Android OS isn't quite as forgiving to wireless beginners as the iPhone is. Setting up your e-mail, contacts and calendar on Android is a breeze (if you're all about Gmail, that is), but when it comes to, say, your music and videos, you're on your own with Android, which lacks an official media syncing client for the desktop. With the iPhone, you do all your syncing on easy-to-use iTunes, which also lets you manage your e-mail accounts, contacts, apps and photos. Then again, you can only use iTunes for syncing the iPhone, while Android users have a variety of third-party options. That's just one example, but in general, Android gives you more options and choices about how you manage your phone and your mobile content — great for experienced and advanced users, but potentially intimating for new mobiles. On the other hand, while beginners might appreciate the (usually) smooth, user-friendly experience that Apple has devised for the iPhone, advanced users may (and often do) get frustrated by Apple's tight control over what they can and can't do on the iPhone. It's a trade-off, plain and simple, and your choice of platform depends on what's right for you.

106. Why would someone choose an Android phone over an iPhone?

One reason is that Android phones boast tight integration with Google services like Gmail, Google Calendar, Google Contacts and Google Voice — perfect for anyone who uses

Google for all their e-mails, contacts and events. Indeed, one of the coolest things about Android phones is that the first time you fire one up, you enter your Google user name and password, then all your Google messages, contacts and other info start syncing into your new handset automatically, no desktop syncing needed. Android is also far more open when it comes to applications. Whereas Apple takes a "walled garden" approach to its App Store, Google won't restrict you from installing apps that aren't featured in its official Android Marketplace. iPhone users, on the other hand, must "jailbreak" their phones if they want to install apps that weren't approved by Apple for inclusion in the App Store. Finally, because Android is open to all manufacturers, a wide variety of Android phones are available to choose from — big and small, souped-up and pared-down, some with slide-out keyboards and some that are all-touch screen. Indeed, in the past few months, a new Android phone has debuted practically every week, while we only get a single new iPhone each year.

****
****

Please check this out:

Our other best-selling books are-

500+ Java & J2EE Interview Questions & Answers-Java & J2EE Programming

200+ Frequently Asked Interview Questions & Answers in iOS Development

200 + Frequently Asked Interview Q & A in SQL , PL/SQL, Database Development & Administration

100+ Frequently Asked Interview Questions & Answers in Scala

100+ Frequently Asked Interview Q & A in Swift Programming

100+ Frequently Asked Interview Q & A in Python Programming

100+ Frequently Asked Interview Questions & Answers in Android Development

100+ most Frequently Asked Interview Questions & Answers in Manual Testing

Frequently asked Interview Q & A in Java programming

Frequently Asked Interview Questions & Answers in J2EE

Frequently asked Interview Q & A in Mobile Testing

Frequently asked Interview Q & A in Test Automation-Selenium Testing

www.ingramcontent.com/pod-product-compliance
Lightning Source LLC
Chambersburg PA
CBHW031249050326
40690CB00007B/1029